ENERGY SECTOR STANDARD OF THE PEOPLE'S REPUBLIC OF CHINA

中华人民共和国能源行业标准

Specification for Pit Exploration of Hydropower Projects

水电工程坑探规程

NB/T 10340-2019

Replace DL/T 5050-2010

Chief Development Department: China Renewable Energy Engineering Institute

Approval Department: National Energy Administration of the People's Republic of China

Implementation Date: July 1, 2020

China Water & Power Press

中国水利水电出版社

Beijing 2024

All rights reserved. No part of this publication may be reproduced, stored in a retrieval system, or transmitted in any form or by any means—electronic, mechanical, photocopying, recording or otherwise, without prior written permission of the publisher.

图书在版编目（CIP）数据

水电工程坑探规程：NB/T 10340-2019 = Specification for Pit Exploration of Hydropower Projects（NB/T 10340-2019）：英文 / 国家能源局发布. -- 北京：中国水利水电出版社, 2024. 8. -- ISBN 978-7-5226-2718-2

Ⅰ. TV22-65

中国国家版本馆CIP数据核字第2024DH0830号

ENERGY SECTOR STANDARD
OF THE PEOPLE'S REPUBLIC OF CHINA
中华人民共和国能源行业标准

Specification for Pit Exploration of Hydropower Projects
水电工程坑探规程
NB/T 10340-2019
Replace DL/T 5050-2010
（英文版）

Issued by National Energy Administration of the People's Republic of China
国家能源局　发布
Translation organized by China Renewable Energy Engineering Institute
水电水利规划设计总院　组织翻译
Published by China Water & Power Press
中国水利水电出版社　出版发行
　Tel: (+ 86 10) 68545888　68545874
　sales@mwr.gov.cn
　Account name: China Water & Power Press
　Address: No.1, Yuyuantan Nanlu, Haidian District, Beijing 100038, China
　http://www.waterpub.com.cn
中国水利水电出版社微机排版中心　排版
北京中献拓方科技发展有限公司　印刷
184mm×260mm　16开本　4印张　127千字
2024年8月第1版　2024年8月第1次印刷
Price（定价）：￥650.00

Introduction

This English version is one of China's energy sector standard series in English. Its translation was organized by China Renewable Energy Engineering Institute authorized by National Energy Administration of the People's Republic of China in compliance with relevant procedures and stipulations. This English version was issued by National Energy Administration of the People's Republic of China in Announcement [2023] No. 1 dated February 6, 2023.

This version was translated from the Chinese Standard NB/T 10340-2019, *Specification for Pit Exploration of Hydropower Projects*, published by China Water & Power Press. The copyright is reserved by National Energy Administration of the People's Republic of China. In the event of any discrepancy in the implementation, the Chinese version shall prevail.

Many thanks go to the staff from relevant standard development organizations and those who have provided generous assistance in the translation and review process.

For further improvement of the English version, any comments and suggestions are welcome and should be addressed to:

China Renewable Energy Engineering Institute
No. 2 Beixiaojie, Liupukang, Xicheng District, Beijing 100120, China
Website: www.creei.cn

Translating organization:

POWERCHINA Chengdu Engineering Corporation Limited

Translating staff:

ZHANG Guangxi   LI Fenglin   YI Cong   GU Jiewei

KE Shanjun   LIU Xiaogang

Review panel members:

LIU Xiaofen	POWERCHINA Zhongnan Engineering Corporation Limited
GUO Jie	POWERCHINA Beijing Engineering Corporation Limited
QUE Chunsheng	Senior English Translator
QIAO Peng	POWERCHINA Northwest Engineering Corporation Limited

CHEN Li	POWERCHINA Kunming Engineering Corporation Limited
LI Shan	Chengdu University of Technology
WEI Yingjie	China Renewable Energy Engineering Institute
WANG Huiming	China Renewable Energy Engineering Institute

National Energy Administration of the People's Republic of China

翻译出版说明

本译本为国家能源局委托水电水利规划设计总院按照有关程序和规定，统一组织翻译的能源行业标准英文版系列译本之一。2023年2月6日，国家能源局以2023年第1号公告予以公布。

本译本是根据中国水利水电出版社出版的《水电工程坑探规程》NB/T 10340—2019 翻译的，著作权归国家能源局所有。在使用过程中，如出现异议，以中文版为准。

本译本在翻译和审核过程中，本标准编制单位及编制组有关成员给予了积极协助。

为不断提高本译本的质量，欢迎使用者提出意见和建议，并反馈给水电水利规划设计总院。

地址：北京市西城区六铺炕北小街2号
邮编：100120
网址：www.creei.cn

本译本翻译单位：中国电建集团成都勘测设计研究院有限公司
本译本翻译人员：张光西　李奉霖　易　聪　辜杰为
　　　　　　　　柯善军　刘晓岗

本译本审核人员：

刘小芬　中国电建集团中南勘测设计研究院有限公司
郭　洁　中国电建集团北京勘测设计研究院有限公司
郄春生　英语高级翻译
乔　鹏　中国电建集团西北勘测设计研究院有限公司
陈　砺　中国电建集团昆明勘测设计研究院有限公司
李　山　成都理工大学
魏颖婕　水电水利规划设计总院
王惠明　水电水利规划设计总院

国家能源局

Announcement of National Energy Administration of the People's Republic of China
[2019] No. 8

National Energy Administration of the People's Republic of China has approved and issued 152 energy sector standards including *Code for Operating and Overhauling of Excitation System of Small Hydropower Units* (Attachment 1), and the English version of 39 energy sector standards including *Code for Safe and Civilized Construction of Onshore Wind Power Projects* (Attachment 2).

Attachments: 1. Directory of Sector Standards

                2. Directory of English Version of Sector Standards

National Energy Administration of the People's Republic of China

December 30, 2019

Attachment 1:

Directory of Sector Standards

Serial number	Standard No.	Title	Replaced standard No.	Adopted international standard No.	Approval date	Implementation date
...						
15	NB/T 10340-2019	Specification for Pit Exploration of Hydropower Projects	DL/T 5050-2010		2019-12-30	2020-07-01
...						

Foreword

According to the requirements of Document GNKJ [2016] No. 238 issued by National Energy Administration of the People's Republic of China, "Notice on Releasing the Development and Revision Plan of Energy Sector Standards in 2016", and after extensive investigation and research, summarization of practical experience, consultation of relevant Chinese standards, and wide solicitation of opinions, the drafting group has prepared this specification.

The main technical contents of this specification include: preparation, exploratory adit, exploratory shaft, exploratory pit and trench, exploratory adit under river, acceptance and quality assessment, and occupational health and safety.

The main technical contents revised are as follows:

— Adding the content relating to field reconnaissance and operations program.

— Adding the content relating to carbon dioxide jet rock breaking.

— Adding the section "Tunnelling with Cantilever Roadheader".

— Adding the content relating to upper inclined shaft operation.

— Deleting the content relating to detonating wire detonation network, compression blasting, and throw blasting.

— Deleting the content relating to transmitted power for various distances and wire diameters in Appendix B (informative).

— Deleting Appendix H (informative) Curve Track Gauge Widening and Outer Rail Raising.

— Revising the content relating to exploratory adit under river into a separate chapter.

— Revising the content relating to acceptance and quality assessment.

— Revising the content relating to occupational health and safety into a separate chapter.

National Energy Administration of the People's Republic of China is in charge of the administration of this specification. China Renewable Energy Engineering Institute has proposed this specification and is responsible for its routine management. Energy Sector Standardization Technical Committee on Hydropower Investigation and Design is responsible for the explanation of specific technical contents. Comments and suggestions in implementation of this specification should be addressed to:

China Renewable Energy Engineering Institute
No. 2 Beixiaojie, Liupukang, Xicheng District, Beijing 100120, China

Chief development organization:

POWERCHINA Chengdu Engineering Corporation Limited

Participating development organizations:

Yellow River Engineering Consulting Co., Ltd.

POWERCHINA Zhongnan Engineering Corporation Limited

Chief drafting staff:

XIE Beicheng	GAN Daming	ZHANG Guangxi	XU Jian
XIN Zongcheng	FENG Shengxue	MOU Lianhe	TANG Minghong
HUANG Guokuan	LI Sheng	LI Wenlong	ZHANG Zonggang
LI Yongfeng			

Review panel members:

YANG Jian	WANG Huiming	GUO Yihua	LI Wengang
LI Xuezheng	XIAO Wanchun	LUO Zhigang	XU Weiqiang
YU Shengbing	SU Jingyi	LI Zhiyuan	XU Qiyun
MAO Huibin	ZHAO Zhenqing	XIONG Dengyu	LI Shisheng

Contents

1	**General Provisions**	1
2	**Terms**	2
3	**Basic requirements**	3
4	**Preparation**	4
5	**Exploratory Adit**	7
5.1	General Requirements	7
5.2	Blasthole Arrangement	8
5.3	Drilling	9
5.4	Blasting	11
5.5	Ventilation	15
5.6	Mucking	17
5.7	Support	19
5.8	Drainage and Dewatering	21
5.9	Tunnelling with Cantilever Roadheaders	22
6	**Exploratory Shaft**	24
6.1	General Requirements	24
6.2	Vertical Shaft	24
6.3	Inclined Shaft	29
7	**Exploratory Pit and Trench**	32
8	**Exploratory Adit Under River**	34
8.1	General Requirements	34
8.2	Access Shaft	35
8.3	Exploratory Adit Under River	35
9	**Acceptance and Quality Assessment**	37
9.1	Quality Criteria	37
9.2	Acceptance and Quality Assessment of Pit Exploration	37
10	**Occupational Health and Safety**	42
10.1	Occupational Safety	42
10.2	Occupational Health	43
Appendix A	Original Record of Exploratory Adit and Shaft	45
Appendix B	Powder Factor for Exploratory Adit Tunnelling	46
Appendix C	Maximum Allowable Concentration of Hazardous Substances in Air	47
Appendix D	Acceptance and Quality Assessment of Exploratory Adit and Shaft	48
Explanation of Wording in This Specification		49
List of Quoted Standards		50

1 General Provisions

1.0.1 This specification is formulated with a view to standardizing the technical requirements for pit exploration of hydropower projects to ensure the pit exploration quality and work safety.

1.0.2 This specification is applicable to the pit exploration of hydropower projects.

1.0.3 In addition to this specification, the pit exploration of hydropower projects shall comply with other current relevant standards of China.

2 Terms

2.0.1 pit exploration

method of exploration to ascertain the engineering geological conditions of rock and soil mass by excavating exploratory adits, exploratory shafts, exploratory pits, and exploratory trenches

2.0.2 exploratory adit

tunnel excavated with an included angle of not greater than 6° to the horizontal plane

2.0.3 exploratory shaft

tunnel excavated with an included angle greater than 6° to the horizontal plane and a depth of not less than 3 m, classified as an inclined shaft or a vertical shaft

2.0.4 inclined shaft

exploratory shaft excavated with an included angle of greater than 6° but less than 45° to the horizontal plane

2.0.5 vertical shaft

exploratory shaft excavated with an angle of not less than 45° to the horizontal plane

2.0.6 open caisson

method of vertical shaft excavation by perpendicularly sinking the caisson

2.0.7 exploratory pit

pit with a depth of less than 3 m, vertically excavated from the ground surface

2.0.8 exploratory trench

long narrow channel excavated in the ground

2.0.9 carbon dioxide jet rock-breaking

process of rock breaking by applying to the surrounding rock mass the high pressure, which is generated from the instantaneous liquid-gas phase change of carbon dioxide

3 Basic requirements

3.0.1 The pit exploration operators shall be trained on safety and skills. For special operations, the operators shall have the certification required for the job.

3.0.2 Pit exploration equipment and tools shall be reasonably configured according to the terms of reference (TOR) and operating conditions.

3.0.3 The purchase, transportation, storage, use and destruction of civil explosives shall comply with the current national standard GB 6722, *Safety Regulations for Blasting*.

3.0.4 Measurements shall be carried out timely during pit exploration operations, observation windows shall be reserved for support sections, and the rock wall of pits shall be cleaned after completion.

3.0.5 Records shall be made in the operation process of exploratory adits and exploratory shafts, and the original records should be in accordance with Appendix A of this specification.

3.0.6 The emergency plan for exploratory adit and shaft operations shall be prepared and drilled.

3.0.7 Mucking in pit exploration shall meet the environmental protection requirements.

3.0.8 When new technologies, new materials, new processes and new methods are adopted in the pit exploration, a special technical scheme shall be formulated and the implementation effect shall be verified.

4 Preparation

4.0.1 Before preparing the pit exploration program, a field reconnaissance shall be carried out, which shall include the following:

1. Understand the climate, topography, geomorphy, basic geological data and available exploration data in the operation area.

2. Understand the conditions of transportation, electricity, communication, important buildings, and production and living materials supply in the region of the operation area.

3. Understand local policies and regulations, management regulations on the use of civil explosives, and folk customs.

4. Preliminarily select the route of exploration access and the construction site for temporary facilities.

4.0.2 Before the pit exploration, a program shall be prepared according to the geological survey outline, TOR, and relevant regulations on safety and environmental protection, considering the field reconnaissance data. The pit exploration program should include the following:

1. Project overview.

2. Technical scheme.

3. Schedule.

4. Resources allocation.

5. Quality assurance measures.

6. Occupational health, safety and environmental protection.

4.0.3 After approval of the pit exploration program, a technical briefing shall be made and recorded.

4.0.4 Pit exploration positioning shall be conducted according to the requirements of the TOR, the topography and geomorphy, and local folk customs. Unauthorized repositioning is not allowed.

4.0.5 Access roads for pit exploration shall meet the following requirements:

1. The access road to the exploration position shall be ready before the pit exploration. The access road for equipment handling should not be less than 2 m wide and should have a width no less than 1.5 times the width of the transporting vehicle.

2. The subgrade of the access road shall be stable and solid, and the

bearing capacity of the pavement shall meet the needs of equipment and material transport. Safety protection measures shall be taken and warning signs shall be set up in dangerous areas such as steep slopes and cliffs.

    3   Pedestrian bridges should have a width no less than 2 m, shall be built on a solid and stable foundation, and shall have a bearing capacity meeting the needs of equipment and material transport.

4.0.6 Temporary facilities shall meet the following requirements:

    1   Before operation, appropriate air compressor room, generator room, water supply station and living rooms shall be constructed according to the pit exploration task, workload and site conditions.

    2   The temporary facilities shall not be arranged in the areas prone to natural disasters such as flood, collapse, landslide, and debris flow.

    3   The construction of civil explosive magazines shall comply with the current national standards GB 6722, *Safety Regulations for Blasting*, and GB 50154, *Safety Code for Design of Underground and Earth Covered Magazine of Powders and Explosives*.

4.0.7 The air supply system for pneumatic jackdrills shall meet the following requirements:

    1   The air compressor station should be located near the pithead. When the air supply distance is long, air tanks with safety devices may be provided.

    2   The capacity of air compressor shall be 1.3 times the air volume required for the simultaneous operation of various pneumatic machines and tools and shall be increased as appropriate in extremely cold and oxygen-deficient regions.

    3   The working air pressure of the air compressor shall match that of the jackdrill, and should not be less than 0.5 MPa.

    4   Air supply pipelines shall be well sealed.

4.0.8 The water supply pressure for the jackdrill should not be lower than 0.25 MPa, and should be lower than the working air pressure by 0.1 MPa to 0.2 MPa.

4.0.9 The power supply system shall meet the following requirements:

    1   The power supply transformer or generator should be arranged at the load center of pit exploration.

2 The wire shall be well insulated, and the power distribution box and cabinet shall be provided with leakage protection device.

3 The power supply shall be three-phase 380 V/220 V for exploration equipment, and 36 V or 24 V for lighting. The lighting fixtures should use waterproof lamp holders and shades.

4 The power cables and lighting lines shall be laid separately in a neat manner. The cables and lines should be arranged on the side wall not less than 1.5 m above the floor of exploratory adit.

4.0.10 The materials and diameters of the air and water pipes shall meet the corresponding air and water pressure and flow requirements. The pipelines should be short, straight and smooth, shall be connected tightly, and shall be laid firmly.

4.0.11 The pipelines crossing roads or in the areas rolled by construction machinery should be placed in conduit and buried.

5 Exploratory Adit

5.1 General Requirements

5.1.1 The location and treatment of the adit entrance shall meet the following requirements:

1. The location of the adit entrance shall avoid such areas with natural disaster risks.

2. In the case of excavating in the flood season and rainy season, the adit entrance shall be higher than the flood line.

3. Dangerous stones above the adit entrance and on the slope surface shall be removed, drainage ditches shall be built, and safety fences shall be set when excavating the entrance.

4. Intensive support or forepoling shall be applied to secure the adit entrance excavation in loose deposits or unloaded fractured rock formations.

5. The adit entrance shall be located far away from buildings and densely populated areas. In special circumstances, the entrance shall be provided with protective nets and safety fences; the number of the fence frames shall not be less than 5, and the fence frames should stretch out of the adit at least 3 m.

6. The operations at the adit entrance where there are buildings nearby shall only be carried out with the consent of relevant authorities.

7. Reliable measures shall be taken during operations at the adit entrance above a road to prevent accidents caused by blasting or mucking.

5.1.2 Exploratory adits should adopt a trapezoidal section or an inverted-U section.

5.1.3 The section size for an exploratory adit shall meet the following requirements:

1. The height of the exploratory adit should not be less than 2 m and shall meet the maximum height requirements of the operating equipment and tools in the exploratory adit.

2. The width of the exploratory adit shall be determined according to the maximum width of the equipment and tools in the adit, the width of the sidewalk and the minimum safety clearance between the equipment and the adit wall, and should not be less than 2 m. The minimum safety

clearance should be 200 mm to 250 mm, and the sidewalk width should be 500 mm to 700 mm.

3　The section size should be 2.5 m × 2.5 m to 3 m × 3 m when the length of the exploratory adit exceeds 400 m.

5.1.4　The adit should be uphill, with an average gradient of 0.3 % to 0.7 %.

5.1.5　The drill-and-blast method should be adopted for exploratory adit tunnelling. For medium hard or poorer rocks, a boom-type roadheader may be used. In special cases, the carbon dioxide jet rock-breaking method may be used.

5.1.6　Inspection shall be carried out before working near the heading face, and shall include the following items:

1　Remove dangerous rocks on the roof, wall and heading face of the exploratory adit.

2　Check for misfire and scattered objects such as explosives and detonators.

3　Check the robustness of supports.

4　Check for water inrush or stagnant water on the working area.

5.1.7　During tunnelling in opposite directions, the distance between the two heading faces shall be accurately measured. When the two heading faces are 15 m apart, the tunnelling shall only be continued at one side, and protective facilities and warning signs shall be set at the other side.

5.1.8　The completed exploratory adit shall be marked at the adit entrance, and a safety protective door should be set at the adit entrance.

5.1.9　During tunnelling, a signboard indicating the work status shall be placed at the adit entrance.

5.2　Blasthole Arrangement

5.2.1　The blastholes shall be arranged in accordance with the blasting design.

5.2.2　The arrangement of cutholes for blasting shall meet the following requirements:

1　Burn cut should be used for medium hard or better rocks. Bull holes should be arranged for burn cut. The spacing between the charged hole and the bull hole should be 150 mm to 250 mm. The centerlines of the blastholes shall be parallel to each other, and the blastholes shall have the equal depth.

2 Angle cut should be adopted for medium hard or poorer rocks. The angles between the cutholes and the heading face shall be the same, and the blastholes shall have the equal depth.

5.2.3 Contour holes should be arranged within 50 mm to 150 mm of the contour line of the exploratory adit section, and shall tilt outward to the contour line, the angle should be 5° to 10°, and the hole spacing should be 500 mm to 800 mm. Blastholes shall be arranged at the turning points of the contour line.

5.2.4 Easer holes should be arranged at a roughly equal distance in between the cutholes and the contour holes, and at an equal spacing of 400 mm to 600 mm.

5.2.5 The bottoms of easer holes and contour holes should be on the same plane perpendicular to the axis of the exploratory adit, and cutholes should be 200 mm deeper.

5.3 Drilling

5.3.1 Drilling machines and tools and operating methods shall be selected according to the working conditions, tasks and requirements of the exploratory adit.

5.3.2 The selection of drilling machines and tools shall meet the following requirements:

1 Pneumatic or hydraulic rock drills should be used for drilling. Electric rock drills may be used if pneumatic or hydraulic rock drills are not applicable. Internal combustion engine shall not be used.

2 Hexagonal hollow carbon steel or alloy steel drill rods with an opposite side size of 22 mm and a tail length of 108 mm ± 1 mm should be selected. The drill rod shall be free of obvious bending, the center hole shall be straight through, and the end face of the tail shall be flat.

3 The diameter of cemented carbide drill bit should be 38 mm to 42 mm. Flat bits may be used for relatively intact medium-abrasive rock stratum, as well as for light and medium-sized rock drills; and cross bits and button bits should be selected for the fractured and high-abrasive rock stratum, as well as for rock drills with high impact energy. The bit steel body shall be free of deformation, cracking, or excessive taper deviation, and the cemented carbide is welded on firmly.

5.3.3 Drilling operation shall meet the following requirements:

1 The drilling shall be conducted under the wet condition.

2   Blastholes shall be marked according to the blasting design, and drilled in the designed direction.

3   The drill rod for hole opening should not be too long, and the leg height shall be adjusted to ensure the drill rod remains in the drilling center.

4   Blastholes shall be cleaned up with high pressure air.

5   Drilling in residual holes or wearing gloves to hold the drill rod is not allowed.

5.3.4 Drilling with pneumatic rock drills shall meet the following requirements:

1   The pressure and volume of air and water shall meet the drilling requirements.

2   Air pipes shall be jointed secure and reliable.

3   Air and water supply pipes shall not be folded or squeezed.

5.3.5 Drilling with hydraulic rock drills shall meet the following requirements:

1   The hydraulic oil shall meet the grade and quality requirements specified in the machine specifications.

2   Hydraulic oil level, water pressure and volume shall comply with the working parameter standards of the hydraulic rockdrill.

3   When connecting oil pipes, the air in the pipe shall be drained, and the joints shall be well protected after the oil pipe is dismantled.

4   The oil pipe shall not be folded, twisted or squeezed, and shall be protected from being damaged by smashing, stabbing or abrading.

5   In the segments with considerable water seepage, the electric parts of the hydraulic power unit shall be covered up.

6   The overflow valve switch shall be opened to release the oil pressure before its startup or shutdown.

7   Before the electric motor starts, it shall be inched first to confirm its direction of rotation.

5.3.6 Drilling with electric rock drills shall meet the following requirements:

1   Cables shall be well insulated, and the water pressure and volume shall meet relevant requirements.

2   During drilling operation, empty play shall be prevented. Drilling operation shall stop immediately if the drill rod is stuck, and forced

pull-out by the drill is not allowed.

3 Operators shall wear insulating gloves and footwear.

5.4 Blasting

5.4.1 The blasting operation shall meet the following requirements:

1 The blasting materials and accessories shall be purchased, transported, used and stored in compliance with the relevant provisions of the state.

2 Blasting personnel shall have the certification required for the job.

3 Blasting design shall be carried out specially. The blasting parameters shall be selected reasonably based on the properties of rocks, the performance of explosive, the cross-section size of the exploratory adit, etc. The powder factor for exploratory adit tunnelling may be selected in accordance with Appendix B of this specification.

4 The blasting materials and accessories received by blasting personnel shall not be lost or transferred to others, and shall not be destructed or used for other purposes without permission.

5 After blasting, the misfire and other unsafe factors found shall be timely reported and handled; all the remaining blasting materials and accessories shall be returned to the magazine in time.

5.4.2 Blasting circumstances shall meet the following requirements:

1 The working area shall be free of roof fall or top slough, and the passageways shall be safe and unobstructed.

2 The content of flammable and explosive gas in the air within 20 m from the heading face shall not be more than 1 %, and there shall be no sign of flammable or explosive gas outburst.

3 The heading face shall be free of water inrush and temperature anomalies.

5.4.3 Selection of blasting methods and blasting materials and accessories shall meet the following requirements:

1 The blasting methods and blasting materials and accessories shall be reasonably selected according to the engineering geology and operation conditions.

2 For blasting materials and accessories, the rock explosives should be used; and the millisecond delay detonators should be used for firing.

3 The carbon dioxide jet rock-breaking method may be used in the areas

where the civil explosives are restricted or inappropriate.

5.4.4 Warning and signaling for blasting shall meet the following requirements:

1. For blasting, obvious signs shall be set up around the boundary of the defined warning area and sentries shall be assigned for blast warning.

2. When the early warning signal is issued, site clearing shall begin within the warning area, and sentries shall arrive at the designated location according to the instruction and stick to their posts.

3. After confirming that all personnel and equipment have been evacuated from the alarming area, all sentries shall be in place. The firing signal shall not be sent out until it is ready for safe initiation.

4. After blasting, the "all-clear" signal can be sent out only when it is confirmed safe.

5. All types of signals shall be simple and clear, without any confusion, to ensure being heard or seen clearly by all people in and near the warning area.

5.4.5 The explosive charging shall meet the following requirements:

1. Wooden or bamboo sticks shall be used for explosive charging. The detonator delay period, loaded constitution and charge quantity shall meet the design requirements for the blasting.

2. Before charging, the blasthole shall be checked for depth and angle compliance with the design.

3. When charging, primer shall not be thrown. Once in place, effective measures shall be taken to prevent the primer from being impacted directly by the subsequent cartridges.

4. If clogging occurs during charging, a non-metallic long rod may be used to declog before the primer is put in place. It is not allowed to impact or squeeze the primer after it is in place.

5. The nonel tube or leg wire of electric detonator in the primer shall not be pulled out or pulled hard during charging.

5.4.6 Stemming shall meet the following requirements:

1. The blasthole shall be stemmed according to the blasting design, and blasting without stemming is not allowed.

2. Clay or tamping bags may be used for blasthole stemming, and stones

and flammable materials shall not be used; the stemming length shall not be less than 200 mm, and should be 1/4 to 1/3 of the hole.

3   Stemming materials shall be tamped, but the tamping force shall not be excessive.

4   During tamping, the stemming materials directly contacting the primer shall not be tamped, and the stemming materials shall not impact the primer.

5   During tamping, the leg wire of the detonator shall not be damaged.

5.4.7   Selection of initiation method shall meet the following requirements:

1   Electric detonator should be fired by initiator or power electricity.

2   The initiation of nonel shall be fired by special initiator or detonator.

3   In the circumstances with gas and dust explosion risk, permitted blasting materials and accessories for coal mines shall be used.

5.4.8   Electric firing circuit shall meet the following requirements:

1   For the same firing circuit, the electric detonators shall be from the same manufacturer, in the same batch, and of the same model. The resistance difference between the electric detonators shall not exceed the value given in the specifications.

2   The common electric detonator shall not be used for firing when the stray current is greater than 30 mA or shall not be used within the safety distance of high voltage RF power supply.

3   The source power for firing shall be able to ensure the initiation of all electric detonators in the circuit.

4   The trunk line of the electric firing circuit shall be routed separately with conductor of good insulation, and shall not be crossed by or mixed with others. The continuity and the resistance of the circuit shall be checked by a blasting galvanometer and a blasting bridge, whose working current shall be less than 30 mA. The checking shall be done once a month.

5   After all the blastholes on the heading face are completely charged and all personnel other than operators are evacuated, the electric firing circuit shall be connected in sequence from the heading face to the detonating station.

6   The electric firing circuit shall not be used during thunderstorms.

5.4.9 Nonel initiation system shall meet the following requirements:

1 The nonel initiation system shall be connected in strict conformity with the design. There shall be no dead knots and no joints in the holes. There shall be sufficient distance between adjacent detonators outside the holes.

2 The distance between the blasting cap and the tie-up end of the nonel shall not be less than 15 cm.

3 The nonel should be evenly laid around the blasting cap and shall be tied tightly. The nonel shall be protected from being cut off by shaped jet of detonator initiation.

5.4.10 Postblast inspection shall meet the following requirements:

1 After the blast, inspectors shall wait for 5 min before entering the blast site in the case of no misfire, or 15 min in the case that misfire cannot be confirmed.

2 Ventilation shall be performed immediately after blasting and shall last for at least 15 min. After air in the exploratory adit is confirmed qualified, the operators can enter the heading face.

3 Inspections shall mainly involve checking if there is misfire, roof caving, or unstable rocks, and if support is damaged, or the blasting fume is exhausted.

4 If misfires or other dangers are found, the inspector shall set up warning signs, take security measures, and forbid unauthorized personnel from approaching the site.

5.4.11 Misfires shall be handled according to the following requirements:

1 A warning area shall be defined, and guards shall be set up to prevent unauthorized personnel from entering.

2 Misfires shall be handled by the experienced blaster on the current shift.

3 When the electric firing circuit is used, the power shall be cut off immediately in case of a misfire, and the misfire shall be short-circuited in time. Then, reconnect and fire after the circuit is checked normal.

4 When the nonel initiation system is used, the nonel shall be checked in case of a misfire. The broken or damaged nonel shall be replaced before refiring.

5 If reconnecting cannot be done for refiring, operators shall take out the

stemming in the blast hole and initiate with a primer, or remove the explosive and recover the detonator, or drill a parallel hole for charging and blasting.

5.4.12 Carbon dioxide jet rock-breaking shall meet the following requirements:

1. Before operation, blasting design shall be carried out according to the rock condition and the operating environment. The technical parameters for rock breaking should be determined through field test before blasting design. Blasting design shall mainly include the blasthole spacing, blasthole depth, blasthole diameter and pattern, the choice of expansion pipe, and the initiation system.

2. The diameter of blasthole shall be adapted to the expansion pipe, and should be 1.1 times the diameter of the expansion pipe.

3. The specification of expansion pipe should be selected comprehensively considering the rock strength, operating environment, and blasthole pattern.

4. During operation, the expansion pipe shall be installed firmly, and the initiation system shall be continuous and reliable.

5. The carbon dioxide storage tank shall be far away from the heading face, stored in a safe place and managed by designated personnel.

6. Carbon dioxide filling shall be done by qualified persons. Before filling, the expansion pipe shall be checked for tightness and soundness. The filled pipe shall be handled with care, and its discharge head shall not point to people.

7. Exciter should be fired by the electric initiator, and the circuit shall be connected by special initiation wire.

8. The guarding for carbon dioxide jet rock-breaking should comply with Article 5.4.4 of this specification.

5.5 Ventilation

5.5.1 Ventilation shall be carried out during tunnelling. Forced or natural ventilation may be adopted depending on the distance to the heading face. Axial flow fans should be used for forced ventilation.

5.5.2 Ventilation volume shall be calculated by the following methods, and the largest value shall be taken:

1. Calculated based on 4.0 m^3 fresh air per minute per person at the maximum number of people working at the same time.

2 Calculated with the air velocity no lower than 0.25 m/s as required for dust removal in the adit.

3 Calculated with the allowed maximum concentration of hazardous substances in air near the heading face after blasting in accordance with Appendix C of this specification.

4 Calculated with the exhaust rate of no less than 4.0 m^3/(kW·min) for the internal combustion engines working simultaneously in the adit.

5.5.3 The selection of forced ventilation mode shall meet the following requirements:

1 For exploratory adits with a length shorter than 200 m, blowing or exhausting ventilation should be adopted.

2 For exploratory adits with a length of 200 m or above, combined ventilation should be adopted. The power of exhausting fans shall be 20 % to 25 % greater than that of the blowing fans.

5.5.4 Fan installation shall meet the following requirements:

1 For blowing ventilation, the fan shall be installed at the entrance of the adit and the air inlet shall be more than 3 m away from the entrance of the adit.

2 For exhausting ventilation, the exhaust inlet should be installed about 30 m away from the heading face, and the outlet of the air duct shall be located more than 3 m outside the entrance of the adit .

3 For combined ventilation, the distance between the inlet of the blowing fan and the exhausting fan should be no less than 10 m. The air inlet of the blowing fan shall be located near the entrance of the adit, and the exhaust inlet of the exhausting fan shall be located near the heading face.

5.5.5 Layout of air ducts shall meet the following requirements:

1 The diameter of the air duct shall be determined according to the length of the adit, which should be 300 mm to 500 mm. The fiber rubber or plastic flexible air duct should be used. The rigid duct should be adopted for the air inlet section of the exhausting fan.

2 The hung air duct shall be flat, straight, tight, stable, and smooth, and the rigid air duct should be used for the turning points of the pipeline.

3 The exhaust outlet of blowing ventilation should be located no more than 10 m from the heading face, and the exhaust inlet of exhausting

ventilation should be located no more than 6 m from the heading face.

5.6 Mucking

5.6.1 Appropriate muck loading, transporting, and unloading methods shall be selected according to the length of the adit and the operating conditions. The mucking method shall meet the following requirements:

1. The mucking method may be an appropriate combination of mechanical/manual mucking and tracked/trackless mucking.
2. Mechanical loading and transporting are preferred. Battery locomotive traction should be adopted for mechanical transporting.
3. Manual mucking may be adopted for short adits or adits with no mechanical working conditions.
4. The mucking access shall be paved according to different transportation methods.

5.6.2 For trackless mucking, the exploratory adit floor shall be flat, sharp stones shall be removed, and the floor shall be paved with fine stone.

5.6.3 For tracked mucking, the track laying shall meet the following requirements:

1. The steel track of 8 kg/m should be selected for manual loading, and 12 kg/m or 15 kg/m for mechanical loading. The rail gap shall be narrower than 5 mm and the height difference shall be shorter than 2 mm.
2. The track gauge should be 600 mm. The sleeper spacing should be 800 mm to 1000 mm for trolley, and 400 mm to 600 mm for locomotive.
3. The track slope shall be consistent with the average slope of the exploratory adit, and the height of the two rail surfaces shall be consistent. The straight-line part shall be level and straight, the curve part shall be smooth, and the minimum curvature radius shall be no less than the turning radius of the mine cart.
4. The shunting track shall be set every 80 m to 100 m, and its length shall meet the needs of shunting.

5.6.4 Mucks may be loaded manually or mechanically. The mechanical loading shall meet the following requirements:

1. The cables of loading machine or air ducts shall be hung on the side of the exploratory adit and shall not be dragged on the ground.
2. The loading machine shall be operated and managed by professionally

trained personnel with corresponding certificates.

3   During loading, no one is allowed to stand around the loading machine.

4   After loading, the loading machine shall be parked in a safe place with the power or air supply cut off.

5   The loading machine shall not be used to load overweight stones or other objects.

5.6.5 Manual mucking with mine cart shall meet the following requirements:

1   Unloading wharf shall be firm and reliable with safety protection facilities. After mine cart arrives at the stop, the muck can be unloaded.

2   Every part of the mine cart shall be well lubricated, the brakes shall be responsive and reliable, and the mine cart with brake failure shall not be used.

3   The mine cart should not be over loaded, and the muck shall not drop along the way.

4   A safe distance shall be kept between two running mine carts, and the carts are not allowed to run freely.

5   When passing through a bend or turnout, the mine cart shall slow down to prevent derailment.

5.6.6 Muck transport with locomotives shall meet the following requirements:

1   The speed of a locomotive shall not exceed 10 km/h in the exploratory adit, and shall not exceed 5 km/h when the locomotive is passing through a bend, turnout or section with poor line of sight.

2   When two locomotives travel in the same direction, they shall keep a distance of no less than 60 m and shall slow down.

3   When encountering pedestrians, the locomotive shall slow down and honk to ensure the safety of pedestrians.

4   Muck locomotives shall not be used for carrying persons.

5.6.7 Conspicuous safety signs and protective measures shall be provided at the crossings through which the muck is transported.

5.6.8 The wharf for spoil unloading shall be set up firmly and reliably, and the safety fence and mine cart locking device shall be set.

5.6.9 The muck shall be stockpiled in the designated spoil area, and the selection of the spoil area shall meet the following requirements:

1   The spoil area shall have sufficient capacity.

2   The spoil area shall meet the requirements of soil and water conservation and environmental protection.

5.7 Support

5.7.1 The excavation of unstable adit sections shall be supported in time, and forepoling measures shall be taken for soft and fracture sections with short self-stabilization time. The supporting methods, materials and structures shall be selected according to the surrounding rock geological conditions, construction safety requirements, adit use, service life, and cross-sectional size and shape. The supporting may be wood support, steel support, rock bolt support or their combination.

5.7.2 The supports shall meet the following requirements:

1   The time to support shall be determined according to the geological conditions of the surrounding rock, self-standing time, etc. The supporting shall be completed before the occurrence of harmful relaxation deformation.

2   The supporting facilities close to the working face shall have measures to prevent blasting damage.

3   The supporting facilities shall be inspected regularly, and shall be reinforced or replaced immediately when any damage or deformation is found. The replacement of supporting facilities shall follow the order from the outside to the inside.

4   The supporting work shall be completed in one go.

5   In case of roof fall, the causes shall be found out, and countermeasures shall be developed to deal with it in time.

6   The supporting structure should be in close contact with the surrounding rock or the gap should be filled.

5.7.3 Wood supporting shall meet the following requirements:

1   The wood material shall be dense, and the diameter of the small end shall be no less than 120 mm, rotten or cracked wood material shall not be used.

2   Flat tenons should be used for adit section with high roof pressure and low side pressure; outcrop tenons for adit section with high side pressure and low roof pressure; and skew tenons for adit section with high roof pressure and side pressure.

3 Wood supports shall form an integral structure, joints shall be firm and reliable, and shear supports, horizontal supports and tie rods shall be used between rows.

4 Each set of supports shall be kept on the same plane perpendicular to the adit axis. The gap between the support and the surrounding rock shall be filled tightly with plates, wedges, etc.

5 The supporting columns shall be placed on a flat rock surface, and a cushion beam or closed bottom beam shall be adopted when the ground is relatively soft.

6 Before supporting, the loose rock shall be inspected and treated, and the personnel shall stand in a safe place and shall not work under the loose rock with a short grip tool.

7 The method of "big end up, small end down" should be adopted to stand the column. When using a closed bottom beam, the bottom beam shall be fully embedded in the floor. The supports close to the working face shall be strengthened and protected.

5.7.4 The steel supports shall meet the following requirements:

1 The specification of steel shall be determined according to the surrounding rock geological conditions and the adit cross-sectional size.

2 Supports shall be welded firm and reliable.

3 There shall be a connecting member between the steel frames, and a suitable base plate should be placed below the supporting foot.

5.7.5 The rock bolt supports shall meet the following requirements:

1 The rock bolt supports may be used for blocky rock with poor integrity.

2 The rock bolts should be longer than 1/2 of the height or span of the adit.

3 The type of rock bolt shall be selected according to the surrounding rock conditions and design requirements. The hole position shall be determined with an allowable deviation no greater than 10 cm.

4 When the method of grouting before bolt inserting is used, the hole diameter should be 15 mm larger than the anchor bolt diameter; when the method of grouting after bolt inserting is used, the hole diameter should be 25 mm larger than the anchor bolt diameter for collar grouting, and 40 mm larger than the anchor bolt diameter for downhole grouting.

5 The allowable deviation of cement mortar bolt hole depth shall be ± 50 mm; the hole depth of the expansion shell bolt shall be 50 mm to 100 mm larger than the bolt body length; the hole depth of wedge slotted bolt, resin bolt or cement roll bolt shall not be less than the bolt body length and shall not be 30 mm greater than the bolt body length. The hole depth of friction bolt shall be 50 mm greater than the bolt body length.

6 For rock bolts, the raw material models, specifications, varieties, component quality and technical performance, installation, and quality testing shall comply with the current sector standard DL/T 5181, *Construction Code of Rock Bolt and Shotcrete for Hydropower and Water Resources Project*.

5.7.6 Combined supports may be used when a single type of support, such as wood support, steel support or anchor support, cannot meet the supporting requirements.

5.8 Drainage and Dewatering

5.8.1 Drainage shall be carried out in time when encountering groundwater in the adit exploration work.

5.8.2 Gravity drainage ditches should be used for uphill adits, and water sumps should be dug for pump drainage for downhill adits.

5.8.3 Drainage ditches should be arranged on one side of the adit floor, and the slope gradient should be consistent with that of the adit floor. The adit floor should be slightly tilted to the drainage ditch, the cross section of the ditch should be inverted trapezoid, and the size shall be determined according to the water inflow rate. The drainage ditches shall be frequently cleaned to ensure smooth flow of water.

5.8.4 Pump dewatering shall meet the following requirements:

1 The water pump suction inlet shall have a filter, and the drainage pipeline shall be fixed at the bottom corner of the adit wall.

2 The rated displacement of the water pumps shall exceed 30 % to 50 % of the maximum water inflow rate.

3 When a single pump is used for dewatering, the standby volume shall be 100 %; when two pumps are used for dewatering, the standby volume shall not be less than 50 %.

4 A backup power supply shall be provided for pumps.

5.9 Tunnelling with Cantilever Roadheaders

5.9.1 The installation and testing of the roadheaders shall be performed by designated personnel.

5.9.2 After installation, the roadheaders shall be subjected to idle test first, which shall meet the following requirements.

1 The rotating parts shall be well lubricated, the machine shall have no abnormal sound, the temperature of each part shall be normal, the rotating direction shall be correct, and the meter shall display normally and readings shall conform to the working parameters.

2 The lifting, swiveling and telescoping of the cutting head shall be agile and smooth.

3 The internal and external sprays shall be good and free of water leakage.

4 Hydraulic pipelines shall be connected reliably and free of oil leakage.

5 Loading, transporting and traveling shall be normal, and the control mechanism shall be responsive.

5.9.3 Cutting operation of roadheaders shall meet the following requirements:

1 The roadheader shall be operated by a certified driver. No one is allowed to stand within the cantilever operation range when the roadheader is working.

2 The cutting teeth shall be checked before starting the machine. In the case of fracture or excessive wear, it shall be replaced in time.

3 Before cutting, the cooling water shall be sprayed normally from the nozzle, and the external spray control valve shall be opened first.

4 The cutting head shall approach the working face while rotating, and the operation shall be slow.

5 The cutting depth and width of the cutting head shall be controlled according to the rock hardness. When the cutting head is working, the operating shall be inched to make the cutting head slowly penetrate into the rock.

6 During cutting, if blocking run occurs, the cutting head shall be disengaged or shut down immediately to prevent overload.

7 When operating on the soft adit floor, bearing plates may be laid with a spacing of 1.0 m to 1.5 m.

8   The large rock blocks shall be broken before loading; if the blocks are jammed in the gantry, the operator shall stop cutting immediately for manual crushing, and shall not use the scraper to remove blocks forcibly.

9   When cutting the rock, the back support may be landed before the shovel plate. If necessary, the side support may be landed to enhance stability and reduce vibration.

10 The hydraulic system and water supply pressure shall not be adjusted at will. If an adjustment is required, it shall be operated by an authorized person.

11 Operators shall pay attention to the oil level and oil temperature. When the oil level is lower than the normal working level or the oil temperature exceeds 70 °C, the machine shall be shut down for adding oil or cooling down.

12 In case of abnormal sound or phenomenon, the operator shall shut down the roadheader immediately and find out the cause, and then restart it after troubleshooting.

5.9.4 Operators shall check the switch handles or buttons regularly for agility and reliability and check that the moving and static contacts are in good condition and work normally.

5.9.5 The maintenance and repair shall be performed by professionals, and live maintenance or repair is not allowed.

5.9.6 Ventilation and dust reduction shall be carried out using the onboard ventilation system and spray device. When the onboard device cannot meet the ventilation and dust reduction requirements, the ventilation and dust reduction may be in accordance with Articles 5.5.1 to 5.5.5 of this specification.

5.9.7 When the machine stops working, the shovel plate and cutting head shall be landed.

6 Exploratory Shaft

6.1 General Requirements

6.1.1 When drilling and blasting are required for an exploratory shaft, the drilling shall comply with Articles 5.3.2 to 5.3.6 of this specification, and the blasting shall comply with Articles 5.4.1, 5.4.3 to 5.4.9 and 5.4.11 to 5.4.12 of this specification. Natural ventilation may be adopted for exploratory shafts with a depth of less than 10 m, and blowing ventilation should be adopted for exploratory shafts with a depth of 10 m or higher.

6.1.2 Reliable communications and signals shall be available during operation, and the signals shall be audible and visual.

6.1.3 The shaft head shall be cleaned, leveled and reinforced before the exploratory shaft is excavated, and safety measures such as intercepting (draining) ditches and protective fences shall be provided.

6.1.4 A safety harness shall be worn when working.

6.1.5 A safety guard plate shall be set in the shaft and should be 50 mm to 100 mm thick. The distance between the guard plate and the shaft bottom shall not exceed 2.5 m. The workers shall be under the guard plate during lifting operations.

6.1.6 A water pump shall be used to drain the water if any in the shaft, and the sump shall be set at an appropriate position.

6.1.7 Mechanical mucking should be used for exploratory shafts, and the muck shall be disposed of in accordance with Article 5.6.9 of this specification.

6.1.8 After completion of shaft exploration, the exploratory shaft shall be backfilled or plugged. The exploratory shaft that needs to remain shall be protected with a cover or guardrails.

6.2 Vertical Shaft

6.2.1 The cross section of a vertical shaft shall meet the following requirements:

1. The cross section of a vertical shaft should be rectangular, and its specifications shall be in accordance with Table 6.2.1.

2. When the purpose of exploration is to find out the landslide, the longer side of the shaft section should be perpendicular to the sliding direction.

Table 6.2.1 Specifications of shaft cross-sections

Depth (m)	Net cross section (length × width) (m × m)	Excavation method
< 10	2.0 × 2.0	Manual winch and dewatering pump
≤ 30	3.0 × 2.0	Lifting bucket, dewatering pump, and ladder
≤ 50	3.5 × 2.0	Lifting bucket, dewatering pump, and ladder
≤ 100	4.0 × 2.0	Lifting bucket, dewatering pump, and ladder

6.2.2 The shaft structure and layout shall meet the following requirements:

1 According to the specifications of the shaft section, a wellhead base, a lifting derrick and mucking rails shall be set. Square wood with a side length of 180 mm to 240 mm or circular wood with a diameter of 200 mm to 300 mm should be used for the wellhead base, and the ends shall be at least 50 cm away from the wellhead edges.

2 A safety distance shall be provided between the lifting space and the ladder, and a safety partition should be set.

3 The pipelines for compressed air, water and electricity shall be arranged separately.

6.2.3 The shaft may be excavated manually. Shallow-hole weak blasting should be used in case of a compact stratum or large boulders, and nonel detonators shall be used for initiation. The contour holes shall be spaced as closely as appropriate. The depth of blasthole should be less than 1 m, and the charge quantity should be restricted.

6.2.4 Bench excavation should be adopted for the shaft with a large water inflow. The movable pump pallet should be used for dewatering and avoid interference with the mucking bucket.

6.2.5 According to the stratum conditions, the shaft support methods may be selected from Table 6.2.5.

Table 6.2.5 Shaft support methods

Support methods	Stratum condition
Hanging frame	Bedrock with relatively developed fissures
Inserting plates	Gravelly soil accumulations, bedrock with fractures or fissures developed

6.2.6 Hanging frame support shall meet the following requirements:

1. The base frame at the wellhead shall be buried underground. The diameter of the cross section shall be 1/5 to 1/4 larger than that of the wooden hanging frame. Both ends of the base frame shall be 1.0 m to 1.5 m beyond the well side.

2. The hanging frame shall be made of square wood with a side length of 160 mm to 180 mm or appropriate metal.

3. The hanging frame shall be jointed by flat tenon, and the four corners shall be supported tightly by standing wood. Meanwhile, the back shall be reinforced.

4. The diameter of the tie bar of the hanging frame should be 16 mm to 22 mm, and each pair shall not be less than 8 rebars.

5. The spacing between hanging frames should be 0.7 m to 1.5 m for loose stratum, 2 m to 4 m for fractured rockmass, 4 m to 6 m for relatively intact rockmass, and 8 m to 10 m for intact rockmass. The four corners of the hanging frames shall be embedded into bedrock more than 300 mm.

6. Back plates shall be timely inserted between the hanging frame and the shaft wall, and the gap outside the back plate shall be filled with thatch or other filler.

7. Temporary protective measures shall be taken near the bottom of the well.

6.2.7 Inserting plate support shall meet the following requirements:

1. The inserted board shall be made of fresh and dense wood, with a thickness of 20 mm to 40 mm, a width of 120 mm to 150 mm and a length of 1200 mm to 1600 mm. The lower end shall be sharpened as well.

2. The inserted board should be tilted outward 15° to 20°.

3. The inserted boards shall be tightly combined.

4. During the excavation, the cyclic excavation depth shall be controlled, and the inserted board tip is not allowed to expose. A well frame supporting the inserted board shall be installed for each excavation of 0.5 m to 0.7 m. The well frame shall be made of square wood with a side length of 160 mm to 180 mm or appropriate metal. Adjacent well frames shall be connected and strengthened.

6.2.8 The muck from the shaft shall be lifted out. The lifting operation shall meet the following requirements:

- 1 Hoisting equipment shall be installed firmly. Brake device shall be agile and reliable. Each shift shall inspect and test them before lifting operation.
- 2 Lifting operation shall be conducted by a skilled operator.
- 3 The lifting mechanism shall be equipped with a guide pulley as required, and a depth indicator and limit device as well.
- 4 Muck shall not exceed 4/5 of the capacity of the lifting bucket. The lifting bucket shall not load or unload goods and persons when it hangs in the well or does not stop completely.
- 5 When lifting tools, the tools shall be placed on the bottom of the lifting bucket, the heavy end of a long-handle tool shall be placed downward and be fastened firmly.
- 6 The blasting materials and accessories shall be placed inside the blasting box and secured when lifting.
- 7 A special lifting cage shall be used for personnel transport, and persons in cage are not allowed to put their heads and hands out of the cage.
- 8 The connecting device between the lifting bucket or cage and the wire rope shall not fall off by itself.
- 9 The lifting speed shall be less than 1.0 m/s, and lifting speed for personnel shall be less than 0.5 m/s.
- 10 When the shaft operation is suspended, the wellhead shall be covered with a plate. The wellhead and platform shall be provided with guardrails and safety signs.

6.2.9 The shaft in a loose and weak stratum may be excavated by the open caisson method. The caisson shall be circular in cross section and shall be of a reinforced concrete structure. The caisson shaft should have a depth of 10.0 m to 30.0 m, and an inner diameter of 1.5 m to 3.0 m.

6.2.10 A pilot drift shall be set up when the depth of the open caisson exceeds 10 m. The depth of the pilot drift should be 1/20 greater than that of the open caisson, and the annular gap between the pilot drift and the caisson should be 0.3 m to 0.4 m.

6.2.11 Open caisson shoes shall meet the following requirements:

- 1 The open caisson shoe shall be formed by welding angle steel and steel plates with a thickness of more than 10 mm. The outer diameter of the caisson shoe shall be 100 mm to 200 mm greater than that of the caisson and the height shall be greater than 0.5 m.

2 The blade section of the open caisson shoe shall be bellmouth shaped. The blade angle should be 45° to 60° for gravel-cobble strata, and 25° to 30° for sand-gravel and sandy soil strata.

3 Combined welding and molding shall be carried out on site.

4 The open caisson shoes shall be filled with concrete with a strength class higher than C20. The curing time of concrete before form removal shall be greater than 48 h.

6.2.12 The open caisson shall meet the following requirements:

1 The wall thickness shall be more than 200 mm.

2 The reinforcement ratio for reinforced concrete shall be 4 % to 5 %. Concrete strength class shall be greater than C20. Concrete curing time shall be greater than 48 h before form removal.

6.2.13 Open caisson excavation shall meet the following requirements:

1 A ground concrete cover shall be set up for the open caisson. The cover shall have an area of at least 5 times the cross-sectional area of the open caisson, a thickness greater than 0.2 m and a concrete strength class higher than C15.

2 A movable window shall be set at an appropriate location of the wellbore. It can be opened for observation and geological logging and shall be closed timely after use.

3 The caisson sinking shall be kept vertical during operation. The interval for inclination measurement should be less than 2 m. Corrective measures shall be taken if the center deviation exceeds 100 mm to 200 mm.

4 Gravel pumps may be used for mucking in a sand-cobble stratum, and the suction head shall be less than 4 m. The air-lift reverse circulation method may also be used for mucking, and the submerged depth shall be greater than 4 m.

5 For the gravel-cobble stratum with a particle size not greater than 500 mm, the cone-grab should be used for mucking. For the gravel-cobble stratum with a particle size greater than 500 mm, pump dewatering and manual excavation should be adopted. Blasting should be carried out when encountering large boulders.

6   In the case of dewatering for excavation, the open caisson shall be kept free of water, and dewatering operation shall not stop during suspension of caisson sinking.

6.3 Inclined Shaft

6.3.1 The inclined shaft of less than 30° dip should adopt a trapezoidal section or an arch section. The inclined shaft of greater than 30° dip should adopt a rectangular section. The specifications of an inclined shaft section may be determined in accordance with Table 6.3.1.

Table 6.3.1 specifications of inclined shaft sections

Construction method	Net cross section (height × width) (m × m)
Manual or mechanical excavation and manual winch mucking	2.0 × 2.0
Mechanical excavation and mucking	2.0 × 2.5
Mechanical excavation and tracked mucking	2.0 × 3.0

6.3.2 The electric wires, pipelines and drainage ditch shall be arranged on one side of the inclined shaft and the sidewalk on the other side. The sidewalk shall not be less than 0.5 m wide, and shall be set with steps or a stairway depending on the gradient.

6.3.3 Wellhead excavation shall meet the following requirements:

1   The wellhead of an inclined shaft shall be reliably supported before heading according to the stratum condition.

2   Surface water inflow shall be prevented during operation.

3   Necessary protective facilities shall be set up at the wellhead.

6.3.4 The uphill inclined shaft of greater than 30° dip shall adopt a chute for mucking. The chute type and safety protection measures shall be determined according to the dip of the inclined shaft. The uphill inclined shaft of less than 30° dip should adopt other mechanical mucking methods. The downhill inclined shaft should adopt a wire rope hoist for mucking.

6.3.5 The wire rope hoist mucking for inclined shafts shall meet the following requirements:

1   Refuge holes should be set on the shaft side free of pipelines, wires and

other facilities, at an interval of 50 m to 100 m.

2　Measures shall be taken to prevent the track from sliding when the inclination of the track is greater than 15°.

3　The connecting plates of the tracks shall be bolted.

4　The sloping section of a track shall be connected to the horizontal section through a vertical curve. A reverse slope shall be set between the vertical curve and the straight section, and a stopper shall be set at an appropriate position.

5　The spacing of track sleepers should be 0.6 m to 1.0 m. The sleeper shall be embedded into the ground groove by 1/3 to 2/3 of the thickness. A long sleeper should be laid every 4 m to 8 m along the track and shall extend into the rock groove more than 200 mm.

6　The traction rope shall be aligned with the centerline of the track on the slope section, and ground pulley supports should be provided every 5 m to 10 m. At the turns shall be installed vertical rolls.

7　During lifting operation, the sidewalk is not open for access, and no person is allowed to stand behind the mine cart. The travelling speed of the mine cart should not exceed 1.5 m/s.

8　The lifting equipment shall have a reliable braking system and protective devices for location limit, speed limit, overcurrent, overvoltage and under voltage, and shall be well maintained.

9　In order to prevent the cart slipping, the mine cart shall be equipped with protective devices against disconnection and rope breakage.

10　The wellhead shall be equipped with a cart arrester, safety barrier and security gate.

11　A cart stop shall be set at a certain distance above the working face.

6.3.6 In addition to the requirements of Articles 5.7.2 to 5.7.6 of this specification, the support of an inclined shaft shall meet the following requirements:

1　Longitudinal girder or diagonal bracing shall be strengthened.

2　Splints should be used for connecting the support rods for the inclined shaft of greater than 30° dip.

3　When the dip of the inclined shaft is greater than the stable dip of the floor stratum, the floor shall be equipped with a floor beam.

4　Support columns shall be firmly connected to the bedrock.

6.3.7　A sump may be set in the downhill inclined shaft to collect water for pumping out.

7 Exploratory Pit and Trench

7.0.1 The cross section of an exploratory pit should be rectangular or circular. The opening size of the rectangular section should be 1.5 m × 2.0 m, and the diameter of the circular section should be 1.5 m to 1.8 m.

7.0.2 Pit excavation shall meet the following requirements:

1 The excavation may be conducted by man power with manual winch mucking, or by an excavator.

2 Adobe blasting may be used for pit excavation when encountering a compact formation or large boulders.

7.0.3 The cross section of an exploratory trench should meet the following requirements:

1 The exploratory trench with a depth less than 1 m may be excavated vertically as a rectangular section.

2 The exploratory trench with a depth of 1 m to 3 m should adopt an inverted trapezoidal section. The bottom width should not be less than 0.6 m, and the sidewall inclination should be 60° to 80° generally. For loose soils with a high water content, the sidewall inclination should be less than 55°.

7.0.4 Exploratory trench excavation shall meet the following requirements:

1 The exploratory trench should be excavated by man power or may be by excavator.

2 Weak blasting should be used to excavate loose, brittle or fractured rocks with joints development or tight uncemented or semi-cemented soils.

3 Adobe blasting may be adopted for large boulders and rock blocks with a diameter greater than 0.5 m, homogeneous weathered rocks, frozen and dense soils encountered during trenching.

4 During trenching, loose rocks shall be timely removed, and the exploratory trench wall shall keep smooth. The soil-stone and tools shall not be placed within 1 m from the sidewall top of the trench. The water in the trench shall be drained out 5 m away.

5 It is not allowed to hollow out the bottom of the wall to make it fall down naturally.

6 A safe distance shall be kept when several workers are working in the

trench.

7 The stability of the exploratory trench shall be strictly checked before resuming operations after rain, potential risks shall be eliminated timely, and blind operation is forbidden.

8 The exploratory trenching on slope shall be conducted from top to bottom, and simultaneous operation both at top and bottom is not allowed.

7.0.5 A partition wall with a thickness greater than 1.0 m should be set for a long trench with a depth greater than 1.0 m. Whether support is required shall be evaluated according to the geological conditions. Suitable support measures may be applied for the trench when necessary according to the site conditions.

8 Exploratory Adit Under River

8.1 General Requirements

8.1.1 Field reconnaissance shall be conducted before excavation of the exploratory adit under river in accordance with Article 4.0.1 of this specification, and the hydrological and meteorological data of the site shall also be collected, including:

1. Riverbed across section, water depth, flow rate, flow velocity, historical maximum flood level and flood level with a return period of 20 years.

2. Riverbed drilling exploration data and geological data in the area.

8.1.2 The shape, section size and slope of the access shaft shall be determined according to factors such as the topographic and geological conditions, the buried depth of the exploratory adit under river, the design length, and the construction equipment.

8.1.3 The site layout shall be determined according to the method of excavation and mucking. Temporary facilities such as workshops, equipment and material warehouses, and spoil yards should be arranged on the site.

8.1.4 The entrance selection for the access shaft shall meet the following requirements:

1. The entrance shall be determined comprehensively according to the design requirements, working conditions, flood level, etc. The entrance elevation shall not be lower than the flood level with a return period of 20 years.

2. The entrance area shall avoid dangerous areas such as landslide and deformed body.

8.1.5 The potential water inflow in the exploratory adit under river shall be estimated according to the available hydrogeological data.

8.1.6 Drilling shall comply with Articles 5.3.1 to 5.3.6 of this specification, blasting shall comply with Articles 5.4.1 to 5.4.12 of this specification, ventilation shall comply with Articles 5.5.1 to 5.5.5 of this specification, drainage shall comply with Articles 5.8.1 to 5.8.4 of this specification, and supporting shall comply with Articles 5.7.1 to 5.7.6 of this specification.

8.1.7 Guide devices shall be provided for the tunnelling of the exploratory adit under river.

8.2 Access Shaft

8.2.1 The access shaft may be inclined or vertical, and the inclination of a vertical shaft should be 90°.

8.2.2 For a vertical shaft, the excavation shall comply with Articles 6.2.1 to 6.2.8 of this specification and the following requirements:

1 Before tunnelling, geological exploration holes should be arranged to ascertain the strata conditions. Waterproof measures shall be taken in the water-rich stratum.

2 The shaft excavation shall be conducted in a stepped manner.

8.2.3 For an inclined shaft, the excavation shall comply with Articles 6.3.1 to 6.3.7 of this specification and the following requirements as well:

1 The inclination of the shaft should not be greater than 15°.

2 In the process of tunnelling, advance drilling should be conducted on the roof and both sides of the working face, the depth of borehole should be greater than 5 m, and the number of boreholes should be greater than 3. Water-sealing measures such as advance grouting should be adopted in the shaft section with large water gushing.

8.3 Exploratory Adit Under River

8.3.1 Excavation of the exploratory adit under river shall meet the following provisions:

1 The main sump should be located at the junction between the access shaft and the exploratory adit under river. The sump depth shall not be less than 3 m, and the sump storage capacity shall be calculated based on the hydrogeological test data and the dewatering pump capacity. Safety measures and signs shall be provided at the sump mouth.

2 Depending on the length of the access shaft and the exploratory adit under river, safety or refuge holes should be arranged every 50 m to 100 m along the exploratory adit under river.

3 Advance drilling shall be carried out during excavation of the exploratory adit under river. The number of boreholes shall be determined according to the cross-sectional structure of the exploratory adit, and should not be less than 3. The depth of advance holes shall be determined according to hydrogeological test data, and should not be less than 5 m.

4 During excavating, the permeable layer encountered shall be treated

with measures such as strengthening dewatering and advance grouting.

8.3.2 Advance prediction shall be conducted on the surrounding rock stability and groundwater conditions.

9 Acceptance and Quality Assessment

9.1 Quality Criteria

9.1.1 The depth of the exploratory shaft or the length of the exploratory adit shall meet the design requirements, and the allowable deviation should be ±0.50 m.

9.1.2 The cross-sectional size of the exploratory adit or shaft shall meet the design requirements, and the allowable deviation of side length or diameter should be ±0.20 m.

9.1.3 The tunnelling direction of the exploratory adit or shaft shall meet the design requirements, and the allowable deviation of the center line from the design axis should be ±0.30 m.

9.1.4 There shall be no standing water on the floor of the exploratory adit. The exploratory shaft shall be provided with drainage.

9.1.5 The wall of exploratory adit or shaft shall be cleaned. After cleaning up, the chainage shall be marked every 5 m on the wall at 1 m above the adit floor, and the mark shall be legible and durable.

9.1.6 The unflatness of exploratory adit or shaft should not exceed 0.15 m.

9.1.7 The average slope of the exploratory adit shall be 0.3 % to 0.7 %, and should not be greater than 1.0 % locally. The adit should not be horizontal or downhill. The azimuth and inclination of the inclined shaft shall meet the design requirements.

9.1.8 The exploratory adit and trench shall meet the geological investigation requirements.

9.1.9 The original records shall be timely, accurate, true and complete.

9.2 Acceptance and Quality Assessment of Pit Exploration

9.2.1 The acceptance and quality assessment of exploratory adits and shaft should be performed on a single adit and single shaft basis. After completion, the exploration contractor, exploration manager and geologist in charge shall jointly carry out the site acceptance prior to the demobilization of equipment. The acceptance and quality assessment shall be recorded in accordance with Appendix D of this specification.

9.2.2 The quality of exploratory adits and shafts shall be assessed on a comprehensive basis according to their depth, cross section, tunnelling direction, support, etc. The itemized quality grade and score value should be in accordance with Table 9.2.2.

Table 9.2.2 Quality grade and score of exploratory adit and shaft

Assessment item	Quality grade			
	Excellent	Good	Qualified	Unqualified
Depth	17 - 20	15 - 17	12 - 15	< 12
Cross section	17 - 20	15 - 17	12 - 15	< 12
Direction	25 - 30	21 - 25	18 - 21	< 18
Support	25 - 30	21 - 25	18 - 21	< 18

NOTE Deductions shall be made from the total score if any of the drainage, wall cleaness, wall unflatness, and original records fails to meet their respective requirements.

9.2.3 Scoring for the depth quality of exploratory adit and shaft shall be in accordance with Table 9.2.3.

Table 9.2.3 Scoring for the depth quality of exploratory adit and shaft

S/N	Depth error	Score
1	≤ 0.10 m	17 - 20
2	0.10 m - 0.30 m	15 - 17
3	0.30 m - 0.50 m	12 - 15
4	> 0.50 m	< 12

9.2.4 Scoring for the cross-section quality of exploratory adit and shaft shall be in accordance with Table 9.2.4.

Table 9.2.4 Scoring for the cross-section quality of exploratory adit and shaft

S/N	Cross-sectional size error	Score
1	≤ 0.08 m	17 - 20
2	0.08 m - 0.15 m	15 - 17
3	0.15 m - 0.20 m	12 - 15
4	> 0.20 m	< 12

9.2.5 Scoring for the quality of tunnelling direction of exploratory adit and shaft shall be in accordance with Table 9.2.5.

Table 9.2.5 Scoring for the quality of tunnelling direction of exploratory adit and shaft

S/N	Deviation of centerline from design axis	Score
1	≤ 0.10 m	25 - 30
2	0.10 m - 0.20 m	21 - 25
3	0.20 m - 0.30 m	18 - 21
4	> 0.30 m	< 18

9.2.6 Scoring for the support quality of exploratory adit and shaft shall be in accordance with Table 9.2.6.

Table 9.2.6 Scoring for the support quality of exploratory adit and shaft

S/N	Support quality	Score
1	Support type is reasonable, and the support quality meets the stability requirement	25 - 30
2	Support type is reasonable, and the support quality basically meets the stability requirement	21 - 25
3	Support type is basically reasonable, and the support quality basically meets the stability requirements	18 - 21
4	Support type is basically reasonable, and the support quality meets the minimum stability requirements	< 18

9.2.7 Scoring for the drainage quality of exploratory adit and shaft shall be in accordance with Table 9.2.7.

Table 9.2.7 Scoring for the drainage quality of exploratory adit and shaft

S/N	Drainage	Score
1	The gradient of the exploratory adit meets the requirements, the drainage is smooth, the floor is free of standing water. Shaft drainage meets the requirements	0
2	The gradient of the exploratory adit basically meets the requirements and the drainage is relatively smooth. Shaft drainage basically meets the requirements	-2
3	The gradient of the exploratory adit roughly meets the requirements, and the water is basically drained by gravity. Shaft drainage roughly meets the requirements. There is a small amount of standing water, which does not affect the normal use of adit	-5

Table 9.2.7 *(continued)*

S/N	Drainage	Score
4	The gradient of the exploratory adit meets the minimum requirements, and the water is basically drained by gravity. Shaft drainage meet the minimum requirements. There is standing water, but the adit is still working properly	-8

9.2.8 Scoring for the quality of wall cleaness of exploratory adit and shaft shall be in accordance with Table 9.2.8.

Table 9.2.8 Scoring for the quality of wall cleaness of exploratory adit and shaft

S/N	Wall cleaness	Score
1	The wall of exploratory adit or shaft is clean, and the strata, discontinuities, and other geological phenomena and their boundaries are clearly visible	0
2	The wall of exploratory adit or shaft is basically clean, but there is a small amount of residual rock powder on the wall, and the cumulative length is not more than 5 %	-2
3	The wall of exploratory adit or shaft is basically clean, but there is a small amount of residual rock powder on the wall, and the cumulative length is not more than 10 %	-5
4	The wall of exploratory adit or shaft is basically clean, but there is a small amount of residual rock powder on the wall, and the cumulative length is not more than 15 %	-8

9.2.9 The highest deduction for the quality of the wall unflatness of exploratory adit or shaft is 3 points.

9.2.10 Scoring for the quality of original records shall be in accordance with Table 9.2.10.

Table 9.2.10 Scoring for the quality of original records

S/N	Original records	Score
1	Complete, legible, and binding in order	0
2	The content is basically complete, and the data error and the binding error is not more than 10 %	-1
3	There are a few missing items in the content, and the data error and the binding error is more than 10 %	-2

9.2.11 According to the total score of pit exploration products, the quality of exploratory adit and shaft shall be classified into 4 grades: excellent, good, qualified and unqualified in accordance with Table 9.2.11.

Table 9.2.11　Quality grading of exploratory adit and shaft

Quality grade	Excellent	Good	Qualified	Unqualified
Total score (N)	$N \geq 90$	$90 > N \geq 75$	$75 > N \geq 60$	$N < 60$

9.2.12 If any of the depth, cross section, tunnelling direction and support is assessed unqualified, the quality of the pit exploration shall be downgraded; if two or more of them are assessed unqualified, the quality of the pit exploration shall be assessed unqualified.

10 Occupational Health and Safety

10.1 Occupational Safety

10.1.1 Pit exploration equipment and instruments shall meet the following requirements:

1. The instruments shall be calibrated by competent organizations and be used within the validity period.

2. The equipment and instruments shall be operated in accordance with their instructions.

3. Internal combustion engine shall not be used in enclosed spaces.

10.1.2 The pit exploration area shall meet the following requirements:

1. The layout of pit exploration and temporary facilities shall avoid the areas prone to natural disasters such as floods, landslides, debris flows and flying rocks.

2. Necessary safety protection facilities and obvious safety warning signs shall be set up for the area with high safety risks, and guards shall be posted when necessary.

3. Necessary precautions shall be taken and corresponding first-aid shall be provided for operations in areas threatened by poisonous insects, poisonous snakes and wild animals.

4. When operating in areas with prominent public security risks, adequate security personnel shall be assigned to deal with the situation.

5. Unauthorized personnel shall not enter the working area. The personnel entering the working area shall be informed of the hazards, preventive measures and emergency measures.

10.1.3 The safety of pit exploration operation shall meet the following requirements:

1. Operators working at height shall wear safety ropes and safety belts, and shall not hang safety ropes low. The carry tools and materials shall be tied firmly or put in tool bags and shall not be thrown.

2. Power supply and lighting for operation shall meet the requirements of Article 4.0.9 of this specification, and emergency power supply shall be provided.

3. Charcoal fires for heating is prohibited in enclosed spaces.

4 One person working alone is prohibited.

10.1.4 Hazards shall be identified and evaluated before pit exploration operation. A special operation program shall be prepared for pit exploration in any of the following cases:

1 Potential collapse hazards exist in the thick overburden or unstable rock mass.

2 The exploratory adit is longer than 200 m or the exploratory shaft is deeper than 10 m.

3 Poisonous and harmful substances or rock burst and water gushing might exist in the exploratory adit.

10.1.5 Plugging and backfilling shall be conducted in time upon completion of the pit exploration, and the adits, shafts, pits and trenches that need to remain shall be protected effectively.

10.2 Occupational Health

10.2.1 Pit exploration contractor shall be outfitted with personal protective equipment (PPE) in compliance with relevant regulations and standards of the state and local governments. Operators shall correctly wear PPE.

10.2.2 Air quality control measures shall be taken in pit exploration. During operation, the content of oxygen in the exploratory adit or shaft shall not be less than 20 % by volume, and the maximum allowable concentration of hazardous substances in air shall comply with Appendix C of this specification.

10.2.3 Dedusting at pit exploration operation site shall meet the following requirements:

1 When using wetboring, a hydraulic drill should be used instead of a pneumatic drill.

2 The walls shall be washed regularly and keep clean if geological conditions permit.

3 The muck heap shall be watered before mucking. Each layer of muck shall be loaded and transported after watering thoroughly.

10.2.4 During operation, the average temperature in the adit or shaft should not exceed 28 °C, otherwise, ventilation or cooling measures shall be taken to reduce the temperature. The air velocity shall be adjusted according to the temperature in the exploratory adit and shaft. The relationship between temperature and air velocity in exploratory adit and shaft shall be in accordance with Table 10.2.4.

Table 10.2.4 Relationship between temperature and air velocity in exploratory adit and shaft

Temperature (°C)	< 15	15 - 20	20 - 22	22 - 24	24 - 28
Air velocity (m/s)	< 0.1	< 1.0	> 1.0	> 1.5	> 2.0

10.2.5 The noise at the working face shall not exceed 90 dB. When the noise exceeds 90 dB, noise reduction or other protective measures shall be taken. If the standard is still not reached after taking protective measures, the operators shall reduce their exposure time to noise. Noise and permissible exposure time shall comply with Table 10.2.5.

Table 10.2.5 Noise and permissible exposure time

Noise (dB)	90	93	96	99
Permissible exposure time per working day (h)	8	4	2	1

10.2.6 When the exploratory adit or shaft that has been suspended for a long time needs to resume production or needs to be surveyed, the concentration of oxygen, carbon dioxide, gas and other harmful gases shall be monitored first. When the concentration exceeds the acceptable level, ventilation shall be conducted before the adit and shaft operation.

10.2.7 When working in an environment with radioactive substances, the radioactive rays and gases shall be tested in accordance with the current national standard GBZ 116, *Standard for Controlling Radon and Its Progenies in Underground Space*. When the radioactive substances exceed the acceptable level, measures such as ventilation shall be taken.

Appendix A Original Record of Exploratory Adit and Shaft

Table A Original record of exploratory adit and shaft

Project name:　　　　　　Exploratory adit / shaft No.:　　　Date:

Work content	Operation time			Cross-sectional size:	
	Start time (h: min)	End time (h: min)	Duration (h)	Operation process and problem solving	
Preparations					
Drilling					
Blasting					
Ventilating					
Mucking					
Supporting					
Others					
Operation cycle time (h)					
Chainage at takeover				Chainage at handover	
Blasthole depth (m)				Footage (m)	
Handover crew leader				Takeover crew leader	
Recorded by				Record time	
Main material consumption statistics	Name	Explosive (kg)	Detonator (pcs)	Diesel (kg)	Electricity (kW·h)	Drill bit (pcs/m)
	Quantity					

NOTE The "operation process and problem solving" mainly records the support position, type and quantity as well as the abnormal phenomena such as collapse, water gushing, misfire, rock burst and harmful gas during tunnelling and their treatment.

Appendix B Powder Factor for Exploratory Adit Tunnelling

Table B Powder factor for exploratory adit tunnelling

Cross-sectional area (m^2)	Rock firmness coefficient f				
	2 - 3	4 - 6	8 - 10	12 - 14	15 - 20
	Powder factor (kg/m^3)				
< 4	1.23	1.77	2.48	2.96	3.36
4 - 6	1.05	1.50	2.15	2.64	2.93
6 - 8	0.89	1.28	1.89	2.33	2.59
8 - 10	0.78	1.12	1.69	2.04	2.32
10 - 12	0.72	1.01	1.51	1.90	2.10

NOTE Powder factor in the table are based on the explosive force value of 320 mL. The powder factor for other explosives shall be converted in proportion to the explosive force value.

Appendix C Maximum Allowable Concentration of Hazardous Substances in Air

Table C Maximum allowable concentration of hazardous substances in air

Name	MAC (mg/m^3)	PC-TWA (mg/m^3)	PC-STEL (mg/m^3)
Carbon dioxide (CO_2)	–	9000	18000
Methane (CH_4)	250	–	–
Carbon monoxide (CO)	20	20	30
Nitrogen oxides converted to nitrogen dioxide (NO_2)	–	5	10
Sulfur dioxide (SO_2)	–	5	10
Sulfuretted hydrogen (H_2S)	10	–	–
Aldehydes (acrolein)	0.3	–	–
Dust containing more than 10 % of free SiO_2	–	1	–
Cement dust containing less than 10 % of free SiO_2	–	4	–
Other dusts containing less than 10 % free SiO_2	–	8	–

NOTES

1. Maximum allowable concentration (MAC) refers to the concentration of toxic chemicals that shall not be exceeded at any time during one working day in the workplace.
2. Permissible concentration-time weighted average (PC-TWA) refers to the maximum average concentration of a chemical in air for a normal 8-hour working day and 40-hour week.
3. Permissible concentration-short term exposure limit (PC-STEL) refers to the maximum average concentration to which workers can be exposed for a short period (usually 15 minutes).

Appendix D   Acceptance and Quality Assessment of Exploratory Adit and Shaft

Table D   Acceptance and quality assessment of exploratory adit and shaft

Project name		Location		Exploratory adit/ shaft No.	
Contractor		Cross-sectional size		Commencement date	
Design workload(m)		Actual workload (m)		Completion date	
1. Depth error (Out of 20 points)	Scoring	Reason			
2. Cross section error (Out of 20 points)	Scoring	Reason			
3. Tunnelling direction (Out of 30 points)	Scoring	Reason			
4. Support quality (Out of 30 points)	Scoring	Reason			
5. Drainage (0 to −8 points)	Deduction	Reason			
6. Wall cleaness (0 to −8 points)	Deduction	Reason			
7. Wall flatness (0 to −3 points)	Deduction	Reason			
8. Original record (0 to −2 points)	Deduction	Reason			
Acceptance opinion (In the progress acceptance opinion, if the product is not completed on schedule attributable to the working team, the number of days and the amount of work delayed shall be described) : 1. Description of lithology by section: 2. Progress acceptance: 3. Quality score: Quality grade: 4. Others:					
Exploration person:			Geologist:		
Date:			Date:		
Opinion of the person in charge of exploration: Signature: date:			Opinion of the person in charge of geology: Signature: date:		

Explanation of Wording in This Specification

1 Words used for different degrees of strictness are explained as follows in order to mark the differences in executing the requirements in this specification:

   1) Words denoting a very strict or mandatory requirement:

   "Must" is used for affirmation; "must not" for negation.

   2) Words denoting a strict requirement under normal conditions:

   "Shall" is used for affirmation; "shall not" for negation.

   3) Words denoting a permission of a slight choice or an indication of the most suitable choice when conditions permit:

   "Should" is used for affirmation; "should not" for negation.

   4) "May" is used to express the option available, sometimes with the conditional permit.

2 "Shall meet the requirements of..." or "shall comply with..." is used in this specification to indicate that it is necessary to comply with the requirements stipulated in other relative standards and codes.

List of Quoted Standards

GB 6722, *Safety Regulations for Blasting*

GB 50154, *Safety Code for Design of Underground and Earth Covered Magazine of Powders and Explosives*

GBZ 116, *Standard for Controlling Radon and Its Progenies in Underground Space*

DL/T 5181, *Construction Code of Rock Bolt and Shotcrete for Hydropower and Water Resources Project*